"A must-read for any businessperson who wants to transform a commodity asset into a value-added asset. A surefire way to enhance your business relationships."
Michael W. Hall, Director, Global Consulting Alliances
i2 Technologies

"The tenets of Paul LeBon's book enable you to take control of the out-of-control technology that is today's voicemail. Benefits to the reader are twofold: increased self-empowerment, and greater productivity through the maximum use of time."
Eric J. Sardina, Sales Executive
Ernst & Young

WHAT PEOPLÉ ARE SAYING ABOUT

SCAPE FROM VOICEMAIL HELL

AND ITS AUTHOR, PAUL LEBON.

"*Escape From Voicemail Hell* offers easy survival tips for what can be an obstacle-filled jungle."
Kathleen Kurke, National Practice Leader
Starbridge Group Inc.

"Finally, a 'how-to' book that is set apart from the rest. No pontificating, it's filled with useful, practical examples to help readers from Day One."
Ross Washam, President
Wasco Steel, Inc.

"This powerful tool benefits both the sender and receiver of voicemail. It's so readable — delightful narrative followed by user-friendly suggestions in every chapter."
Barb Taruscio, Performance Consultant
Wilson Learning Worldwide

ESCAPE FROM VOICEMAIL HELL

WRITTEN BY PAUL LEBON

ESCAPE FROM VOICEMAIL HELL

BOOST YOUR PRODUCTIVITY BY MAKING VOICEMAIL WORK FOR YOU

ParLeau Publishing
Highland Village, Texas 75077

www.parleaupublishing.com

"Post-it"® (page xii) is a registered trademark of 3M.

First Edition

Publisher's Cataloging-in-Publication

LeBon, Paul.
    Escape from voicemail hell : boost your productivity by making voicemail work for you / by Paul LeBon. – 1st ed.
      p. cm.
      LCCN: 99-90799
      ISBN: 1-929398-00-X

    1. Voice mail systems. 2. Business communication. I. Title.

HF5541.T4L42 1999   384.6'4
            QBI99-1318

Edited by Sara Karam

Cover and book design by Vincent LoPresti, Dallas, TX

Cover photography by Bill Couture, Dallas, TX

Typesetting by Dawn Cawood, Dawn Graphics, Plano, TX

Author photo by Jason Marzini, Exposure 1, Woonsocket, RI

Dedication

To my wife and partner, Susan, for the love, patience, and encouragement she provided in getting this book to print, and the love and support she provides on a daily basis.

To my sons, Justin and Jared, for the pride and joy they bring to my life, and for continually reminding me not to take myself too seriously.

To my mother, Lillian LeBon, and mother-in-law, Theresa Ivy, for their love and encouragement.

In Special Honor

My father was the son of immigrants who left school in the seventh grade during the Great Depression to work with the WPA and later in textile mills. When my mother was disabled in a factory accident, he worked multiple jobs to feed and clothe his five children, refusing to accept handouts or assistance.

Pushing me to "go to college and make something of myself," his constant refrain was, "Work with your head, not with your back." Those words of encouragement may have been the forerunner of today's motto, "work smarter, not harder".

It is in my father's spirit and counsel, that I offer this book in order that others may find ways to work smarter and boost their productivity.

Narcisse J. LeBon
1917-1997

Table Of Contents

Acknowledgements

First of all, I would like to thank the Good Lord for giving me the inspiration to write this book, and for showing his sense of humor by not claiming copyright infringement when I was inspired to use the term *Ten Commandments of Voicemail*.

On a more worldly note, I would like to thank the many clients, colleagues, and business associates who through the years have worked with me and provided me with inspiration. I especially thank those who came back and told me how beneficial the principles now contained in this book were to them.

Especially, I would like to recognize Judy Kelner, Michael Hall, Betty Tapella, Rusty Washam, Lisa Cass, Eric Sardina, Karen Dugan, Dennis Mankin, Barb Taruscio, Patricia Monbouquette, Warren Bischoff, Kathleen Kurke, David Levy, Thomas Koven, Jane Blinde, and Kevin O'Donnell for their friendship, support, and encouragement. Lest I forget, Sam Davila, my barber who has had the patience to listen to it all.

A special word of thanks to my "German daughter" Steffi Meyer and her family for the wonderful hospitality and respite they provided in June 1999.

I also want to mention the special kids who brighten my life throughout the year, the members of my Hurricanes boys soccer team, and the children at David G. Burnet Elementary School in Dallas.

The concepts would not have been put into this beautifully produced finished version without the effort and direction of Vincent LoPresti, the creative genius who designed the outside "packaging" as well as the interior layout and design. He is a great person and a wonderful loving father to Ivey.

Bill Couture and Jason Marzini contributed their photographic talents, Bill on the front cover and Jason on the author's photo on the back cover, shot by the surf at Hazard Rock.

Especially, my sincerest thanks go to the person responsible for editing my sometimes wordy or cluttered material, Sara Karam, a young lady whose future is shining bright with limitless possibilities.

Foreword

On March 10, 1876, Alexander Graham Bell made history by calling his assistant Mr. Watson in the next room on the telephone and speaking those famous words, "Mr. Watson, come here. I want you."

On November 13, 1900, a Danish inventor named Valdemar Poulsen was granted a patent by the U.S. Patent Office for his invention known as the magnetic tape recorder.

These two technologies were merged in the late 20th century in an effort to give people a means of communicating more efficiently. This technological merger is known as voicemail. Unfortunately, though, many people have never learned to use voicemail properly. They constantly frustrate others by leaving brief messages such as, "Call me, I need to talk to you," which is reminiscent of Mr. Bell's call to Watson, or they drone on and on with an endless message.

Finally, this book offers a solution to help people use their voicemail effectively, and I hope you learn from it and use it as a tool to be more productive. — Paul

Introduction

In my work as a productivity consultant, I will usually ask clients about the challenges they face in their day-to-day work lives. One of the most frequent responses I hear is dealing with or getting around voicemail, or being able to talk live — ear-to-ear — on the telephone with people to get things done.

What an incredible communications medium the telephone is. Within seconds, you can transmit your voice and the message you want to impart to any part of our global village. U.S. presidents have even made telephone calls from the Oval Office to astronauts orbiting space or sitting on the Moon.

In their never-ending quest to improve and enhance our communications capabilities, the telecommunications industry developed a technology to communicate with others via telephone even when those we are trying to reach are unavailable. This technology is commonly known as voicemail. Voicemail has become so prolific that many of us now have it on our home and mobile telephones as well.

In the corporate world, however, people have come to develop a love/hate relationship with voicemail. Some people have realized that voicemail is a great way to hide from people, avoid issues, and put off work. This is the "love" side of the relationship.

Many people see voicemail as a cause of great frustration.

They find themselves calling people and listening to senseless blabber on the person's outgoing message, or returning to their work environment to find they have eight voicemail messages, all saying, "Call me, I need to discuss something very important!"

They return the calls, only to listen again to a long-winded outgoing greeting, advise the person that they are returning their call, and then find themselves in an endless game of telephone tag. I have been told that some of these games of tag go on for weeks! Over the course of time, people develop a deep burning frustration at their inability to communicate with the other party and hence they blame it on the technology they love to hate — "that darned voicemail". They find themselves burning up in *Voicemail Hell*.

I came to realize after a good bit of informal research how this burning frustration and hatred developed. The telecommunications companies who develop and install voicemail systems are true geniuses of technology. The glitch is that, in reality, they are not communications companies, but *technology* companies. These are engineering-driven companies who market a technology that is essentially a soft skill product.

Witness the manner in which they teach people how to use voicemail. Companies that implement a voicemail system usually distribute some type of an engineering flowchart — complete with boxes, lines and arrows. Press this button to retrieve your messages, this button to record your greeting, etc.

Over the past several years, I have developed, in conjunction with client audiences, some principals for effectively using voicemail to communicate productively. Today, I swear by the reliability of voicemail and its contribution to my ability to be more effective and productive, as do many clients. We have learned how efficient and dependable voicemail can be. Since most people's voicemail will pick up calls 24 hours a day, I often leave messages in the evening or early morning hours from the quiet of my home office.

When I place a call and another individual answers and offers to take a message, I generally ask to be put into the voicemail of the person I am calling. While the person answering may be well intentioned, they are likely to pass on a brief edited version of my information. By going into voicemail and leaving the message myself, I ensure that the person I called will get an accurate and complete message, all delivered in my own voice with a positive tone and enthusiasm!

Many clients tell me how much more productive they become when they manage their telephone communications by effectively using voicemail; it is simply a matter of viewing voicemail as a productivity tool rather than a source of constant frustration.

One client in computer mainframe sales told me of the success she had when she stopped viewing voicemail as a source of frustration and began to view it as a productivity tool. She was able to close an additional 30% in sales of add-ons and upgrades to her existing customer base without ever making ear-to-ear contact! She did it all with voicemail.

High-impact voicemail messages help another client who sells industrial supplies successfully prospect for new business. He leaves voicemail messages for prospective buyers late in the evening, so that the buyers receive his message when they first come into their office in the morning.

Another client who was having difficulty leaving the office at the end of the workday realized that he could leave work, have dinner with his family, and work his voicemail early the next morning from home so that his telephone correspondence would all be returned before the start of the business day.

Several two-career couples have regularly used voicemail to communicate reminder messages to each other about household and family matters, as well as getting into the habit of leaving each other sweet, romantic messages each day.

These principles, which I call the *Ten Commandments of Voicemail,* are simple, common sense ways to organize and structure your telephone communications so that you can reduce your anxiety and boost your productivity.

Occasionally you will deal with people who just "won't play". Much as you try to communicate with them in an effective manner, they will ignore you, fail to return calls, leave you vague messages, and exhibit other traits of poor business etiquette. Do not dismay. The problem is not yours, it is theirs! Never allow these people to frustrate you, because you will find that they are truly the exception to the rule, and it is not your responsibility to teach them proper business etiquette.

We also hear a great deal today about the internet and the advent of "e-commerce". While this is certainly a growing segment of our economy and one that will continue to grow, there is still a great deal of "v-commerce" — voice commerce — that will always take place. This book is especially helpful to those who work in areas of customer contact.

How To Use This Book As A Tool To Boost Your Productivity

This book is not a "one-size-fits-all". In reading this book, you will learn many principles and techniques that can help you communicate more effectively via voicemail. There are also examples of the principles in action. I hope you find all the suggestions worthwhile. Yet even if you only put a few into practice, this book will still be a good investment.

In my consulting, I often work with industry veterans who have learned many useful skills in their careers. I counsel them to pick what they like in my presentation, add it to their repertoire, and enhance their skill sets. My favorite expression is, "If all you have is a hammer, everything looks like a nail." Hopefully in this material you will find additional tools to add to your communications toolbox.

So first, read this book cover to cover. Let the ideas and principles sink in a bit. After a couple of days, read it again. Use a highlighter pen and some 3M Post-It Flags to mark the pages you find most helpful to you. Write out several versions of a greeting for your voicemail. Practice messages you might send to business contacts.

Finally, keep the book handy! It is sized just right to fit under most telephones. Sticking it on a shelf or burying it in a drawer somewhere will diminish its long-term value. Refer to it regularly, and hopefully it will give you new skills to communicate effectively and boost your productivity. — Paul

1
Chapter

What Exactly Is Voicemail Hell?

Technology . . . is a queer thing. It brings you great gifts with one hand, and it stabs you in the back with the other.

— C.P. Snow

We live in a world where technology has made our lives easier and our work environment more productive. The telephone ranks as probably one of the greatest inventions of all time because its advent and subsequent enhancements in technology now allow us to communicate across the globe instantly.

Over the past two decades, the telecommunications industry has dedicated itself to enhancing our ability to communicate faster and more efficiently with various breakthroughs in technology. One of the advances in telecommunications technology developed over this period of time has been voicemail.

A technology that today is as ubiquitous as the telephone itself was viewed skeptically fifteen years ago as a path to depersonalizing individual contact. In the mid-1980's we had a magnetic tape answering machine at home and people to accept calls and take messages for us in the workplace. Today, most of us have voicemail at work as well as home now that local telephone service providers are offering it as a feature option. We also have voicemail on our mobile telephones. In fact, some of us feel we have too many voicemail boxes to monitor!

Worldwide, there are over 200 million voicemail subscribers, and the number is growing. Some people predict that e-mail will replace the telephone as a means of communication. I disagree. While e-mail has its benefits, there is still no greater tool that any one of us possesses to persuade, inform, soothe, cajole, or romance — than the extraordinary sound of our own voice. In addition, voicemail is more convenient to access, from anywhere, at anytime, with nothing more than a working telephone required.

Clients also tell me that a voicemail message has a greater imperative to it than an e-mail message. With the proliferation of spam e-mail, it is also getting increasingly difficult to weed out the junk from the serious e-mail messages.

Yet we find ourselves in a love/hate relationship with voicemail. We love it when we can use it to screen our calls, hide from others, and appear to be busy and consumed with work so that to stop and do something like answer the telephone is impossible.

On the other side of this relationship, there is the feeling we get when we call someone with an urgent need for information — a deadline to meet, a boss breathing down our neck with a request, a quick chat just to finalize details of an agreement — and our important call rolls over to voicemail. These are the times when we get upset, angry, frustrated, and a strong burning feeling develops deep down inside of us. This is *Voicemail Hell!*

Companies in the telecommunications industry are doing wonderful things for the world. However, when they developed the technology known as voicemail, they approached the implementation of voicemail systems from an engineering and design standpoint, not from a *communications* standpoint. The only guide or instructions anyone has usually been given when they become new voicemail users is a system access flow chart, a sample of which is depicted on page 5.

While some people have become extremely adept at maneuvering within their voicemail system by studying these flowcharts, they have not enhanced their ability to communicate more effectively.

Now, if an individual goes out and purchases a $5,000 camera and reads the instruction manual, learning how to load film, read the light meter, and set the aperture settings, does this make the person a great photographer? Of course it does not. It takes practice and education in photography techniques, as well as a certain artistic style, to create photos that convey a sense of beauty and artistry.

Likewise, it takes practice and education to communicate effectively and boost your productivity through the use of voicemail. Over the past decade in seminars and workshops, I have often confronted these challenges with participants. In Chapter 2, you will find what has developed over time as the *Ten Commandments of Voicemail*. Subsequent chapters expand on each commandment, practical examples are given, and guidelines are provided so that you may better understand how to practice these principles.

There is no question that the way to avoid burning up in *Voicemail Hell* is to abide by the *Ten Commandments of Voicemail*.

Typical Voicemail Instructional Chart

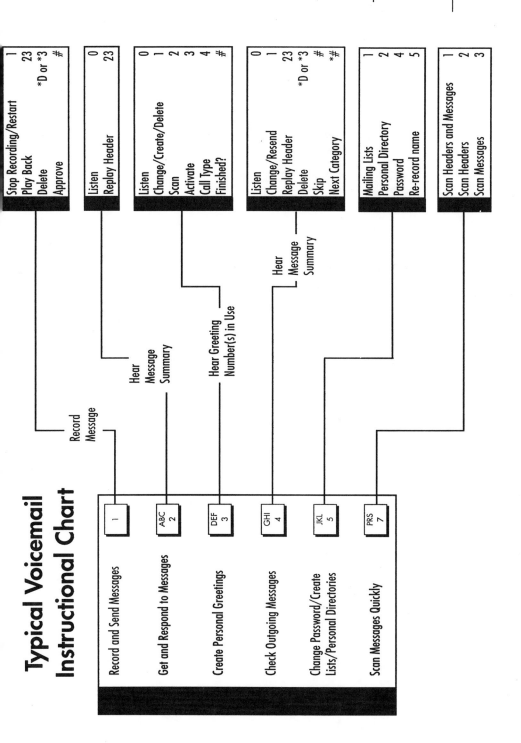

Record and Send Messages [1]

Record Message

Stop Recording/Restart	1
Play Back	23
Delete	*D or *3
Approve	#

Get and Respond to Messages [ABC 2]

Hear Message Summary

| Listen | 0 |
| Replay Header | 23 |

Create Personal Greetings [DEF 3]

Hear Greeting Number(s) in Use

Listen	0
Change/Create/Delete	1
Scan	2
Activate	3
Call Type	4
Finished?	#

Check Outgoing Messages [GHI 4]

Hear Message Summary

Listen	0
Change/Resend	1
Replay Header	23
Delete	*D or *3
Skip	#
Next Category	*#

Change Password/Create Lists/Personal Directories [JKL 5]

Mailing Lists	1
Personal Directory	2
Password	4
Re-record name	5

Scan Messages Quickly [PRS 7]

Scan Headers and Messages	1
Scan Headers	2
Scan Messages	3

2
Chapter

The Ten Commandments Of Voicemail

Excellence demands that you be better than yourself.

—Ted Engstrom

I. Present a clear and concise greeting on your voicemail, including a "call for action".

II. Utilize voicemail to effectively manage and prioritize your callbacks and your time.

III. Plan in advance of calling so that your message is informative and effective.

IV.   Project positive enthusiasm and energy in the messages you leave for others.

V.   Train others to communicate effectively with you via voicemail to eliminate "telephone tag".

VI.   Use internal voicemail and group broadcasts to maximize communications.

VII. Take advantage of the 24x7 capabilities of voicemail.

VIII. "Wow" others via voicemail and build stronger relationships.

IX.   Harness the power of voicemail to prospect for new customers.

X.   Manage and enhance your personal life with voicemail.

3
Chapter

First Commandment:
Present A Clear And Concise Greeting On
Your Voicemail, Including A "Call For Action"

Keep It Short and Simple.
  —The KISS Rule

I am generally amused and bewildered by the amount of information people put on their voicemail greeting. This may include but not be limited to their name, job title, company, location, day of the week, an apology for missing your call, and a comment about being on the phone, or away from their desk — and the list goes on and on.

Then comes some innocuous statement to the effect that your call is very important, followed by a request that you leave a date, time, message, and your number. They generally close with a promise to get back to you as soon as possible. I've also heard some voicemail greetings that included college fight songs, sports statistics, Hollywood trivia, political commentary, celebrity voice impersonations, and a myriad of other unprofessional noise.

The usual reply left for such a greeting is a quick recitation of the caller's name, telephone number, and occasionally a statement about the purpose of the call. [*"Hi Mike, this is Kevin Morris, please call me at 555-0727 to talk about the new product specs. Thanks."*]

In a business environment, it is imperative to give a professional impression, especially a *first* impression. Often you may receive calls from people in remote locations with whom you may never have face-to-face contact. These people judge you and decide whether they want to have a business relationship with you based on the tone, character, and professionalism of your telephonic communications.

To construct a message which is clear, concise, and calls for action, begin with the premise that people who call you know who you are, which company you work for, and that you are unable to answer the telephone. After all, when you answer live, you merely use your name, right? Hence, your voicemail greeting should merely open with your name, and perhaps your company name, followed by an acknowledgement for missing the call. [*"Hello, this is Bob Watson with Harvest Foods, I'm sorry to miss your call."*]

If your schedule fluctuates from day to day, or you are in and out of the office, stating the date and your location for that day is certainly appropriate. ["*Hello, this is Linda Graves, today is September 15 and I will be in (or out of) the office today.*"] If you do note the date on your greeting, however, *always* be certain to change it daily. Nothing presents an unfavorable impression like a three-day-old greeting.

In addition, your system probably time-stamps when the message was received, and even if it does not, is this really important? When I hear, "Please leave the date and time that you called," I begin to wonder if my call will be returned this month!

Next, include a "call for action" so that you can make the callback prepared and ready to move the dialogue forward. ["*Leave me a message with enough detail so that when I return the call, I can have the information you are looking for.*"]

As you will learn in later chapters, prioritizing your callbacks and even making them during off hours can really boost your productivity, so you may want to inquire as to available times to call back. ["*Let me know the best time to reach you, and I'll do my best to get back to you when you are available.*"] ["*If you have 24-hour voicemail, let me know because I sometimes make callbacks after hours.*"]

Finally, if your work or position is time sensitive or mission critical, you may also refer your callers to a third party who can offer assistance. ["*If you require immediate assistance, please press '0' and ask to be transferred to my colleague, Terry Brown.*"]

Presenting a clear, concise, and well-structured greeting on your voicemail that includes a call for action gives your callers the impression that you are a professional, organized individual who is ready to do business.

Here is an example of how a typical wordy, ineffective voicemail greeting sounds.

"Hello, you've reached the voicemail box of Mike Holly, regional sales representative for Creative Industries. I'm either on the phone or away from my office right now, but your call is important to me. Please leave a message with the date and time that you called, and I'll get back to you as soon as I can. Thanks for calling, and have a nice day."

The following examples may assist you in scripting your own professional voicemail greeting.

Examples Of Well-Structured Greetings With A "Call for Action"

"Hello, you've reached the office of Susan Olson. I'm sorry I've missed your call. Leave me a detailed message so that I can get back to you with the information or answers you need. Let me know the best time to reach you and I'll try to get back to you then. Thank you for your call."

"Hi, this is Maureen and you've landed in my voicemail. Leave me a request for action, I'll take action and get back to you. Let me know if you have voicemail after hours, because I often return calls in the early evening. Thanks, and I'll get back to you soon."

"Hello, you've reached the voicemail of Mark Hall. Today is May 11th and I am in the office but will be in meetings most of the day. Leave me a detailed message and I'll respond to your call at my earliest opportunity. If you have an urgent need, you can dial 'O' and ask to be transferred to my administrative assistant, Charlene. Thanks, and have a great day."

"Hello, you have reached the order fulfillment group's voicemail box. We apologize that we are unable to take your call at this time. Please leave a message with your name, account number, order quantity by SKU number, and telephone number, and someone will get back to you shortly to confirm the order. Thank you for your business."

"You've reached Justin Williams' voicemail and I'm sorry to miss your call. Please leave me a message with the specifics of what information I will need when I return your call. I generally return my calls between 3:30 and 4:30 each afternoon. Please be sure you leave your return number, and state it slowly. Thank you."

"Hello, this is Jared Tyler and today is Tuesday, October 18th. I will be traveling out of town for the day and will not have an opportunity to check my messages until this evening. Leave me a detailed message and I will respond to it then. If you need assistance immediately, please press 'Pound' and the letter 'T', then dial extension 1029 and one of my support staff will be able to assist you."

"Hello, you've reached Brighid McDonald. I regret that I've missed your call. Please leave me an explicit message and I will get back to you. Please be sure and leave your number as well as the best time to reach you and I'll do my best to get back to you then. Thank you."

"Hi, this is Mitch Wickman and unfortunately I cannot take your call at this time. If you would please leave a detailed message, I can get back to you later today to address whatever is on your mind. Speak clearly, and give me as much detail as possible. Talk to you soon."

"Hello, this is Jennifer Heller. I'm sorry to miss your call. I will return it later today with responses to any questions you leave in a message. Please be explicit and I will call you back with answers. Thank you."

Greeting Structure Guidelines

1. Open with your name, company name (if appropriate), and an acknowledgement that you have missed the call.

2. If your schedule fluctuates, state the date and your schedule for the day, and update your greeting daily.

3. Pose a "call for action" which will allow you to return the call prepared to move the dialogue forward.

4. Ask the caller to suggest an appropriate callback time.

5. Inquire as to the time capabilities of their voicemail (24 hours, 7days).

6. Include the name and extension number of a back-up person to contact if the call requires immediate action.

7. If your system offers the capability, pre-record and store several greetings for variety and alternate them periodically.

4
Chapter

Second Commandment:
Utilize Voicemail To Effectively Manage And
Prioritize Your Callbacks And Your Time

We must use time as a tool, not as a couch.

— John F. Kennedy

Yes, time is a tool, but as we know some people use time and voicemail as a couch to delay, defer and otherwise avoid having to fulfill the responsibilities of their job.

One of the keys to successful time management is to prioritize your activities. Using the A-B-C method, you can clearly set your daily schedule by giving highest priority to those activities (A's) which give you the highest return on your time investment.

Another key to successfully managing your time is to minimize interruptions. For many of us, telephone calls — while important to our work and business — can be interruptions because they arrive unannounced and usually while we are engrossed in some important task.

Allowing your calls to roll to voicemail is not necessarily a bad thing. This allows you to manage the calls, instead of allowing the calls to manage you. In a situation where you are working against a critical deadline, this makes perfect sense.

Always dedicate a certain amount of time on your calendar each day for handling callbacks. You can advise callers in your greeting that you will be returning calls between 3:00 and 4:30 p.m., for example.

When you listen to your messages, take immediate action with them. Take notes on the caller, issue, and callback number then delete the call. Forward messages to another party who can better deal with the issue if appropriate. This will help alleviate the problem of callers hearing that your voicemail box is full, thereby creating an impression that you are not a well-organized individual.

For those who work on the outside, a mobile phone can be valuable in managing messages and callbacks. Pull off the road to take a break from traffic and work your voicemail. On a nice day, drive to a park or open space for a lunch break and sit on a bench and make your callbacks. Leave voicemails while waiting for client appointments or between calls.

Examples Of Prioritized Callbacks

"Hi Bill, my name is Jack Schaeffer with Andor Corporation and we are looking for vendors to respond to an RFQ for some raw materials we will be needing to produce a new product we are rolling out later this year. Could you please call me at 555-8135 so that we can discuss this? Thank you."

"Hello Bill, this is Kendall Brown. We've run short of supply for our current production run and we need to get 15,000 additional units shipped to us by the middle of next week. Our purchase order number is 102956. Could you get back to me and confirm availability and let me know when we might expect delivery?"

"Mr. Simmons, this is Robert in HR. I need to speak with you regarding a few questions I have about the sales assistant candidate before I can initiate the new hire process."

"Hey Bill, this is Louise. We have a meeting set-up tomorrow to review all bids. I need to know the number of units per case and the tare weight per case of the 5509's. Please get back to me before 10:30 a.m. tomorrow, as I need this information for the meeting."

Clearly the second message is an "A" — a $100 callback. It represents sales and income.

The third and fourth messages are both likely "B" calls. The difference is that one requires live ear-to-ear contact (Robert in HR), while the latter can be responded to anytime, even after hours.

The first message is likely a "C" — a prospective new customer, looking ahead long term. Not that this callback should not be treated seriously, it can merely be deferred to a more convenient time — one that is more convenient for you.

Callback Management Guidelines

1. Review your messages periodically throughout the day.

2. If possible, forward to an individual who can take more appropriate action.

3. Make a written list of callers and subject matter, along with the type of information you need to collect before returning the call.

4. Prioritize the callbacks using the A-B-C method, based on return on the time investment needed to make the callback. (Think of A's as $100 calls, B's as $50 calls, and C's as $10 calls.)

5. Dedicate a certain amount of time in your schedule throughout the day to make callbacks.

6. Determine which callbacks can be deferred until after-hours, early morning, or even the weekend!

7. Make your callbacks and be prepared to share the information the caller was seeking.

8. Keep in mind that returning a callback after hours and leaving it on a person's voicemail can reduce a 15 minute ear-to-ear conversation down to a mere 30 second information exchange. (Learn more about this in Chapter 10.)

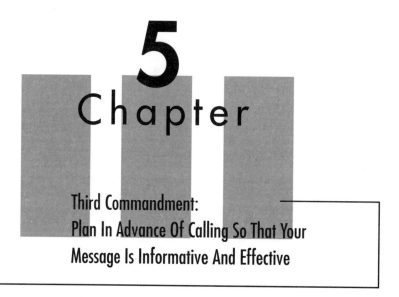

5
Chapter

Third Commandment:
Plan In Advance Of Calling So That Your
Message Is Informative And Effective

Many attempts to communicate are nullified by saying too much.

— Robert Greenleaf

Listening to one's voicemail messages can be a very frustrating and infuriating experience. *"Hello Mary, it's Frank Martin with Cerromar Corporation. Please call me back at 212-555-9#3#."* The caller spoke so fast that poor Mary had to replay the message five times to understand the callback number.

Equally frustrating are callers who leave their name, date, time, and callback number with no stated purpose for the call. They merely end by asking you to call them back. This type of message is a "Z" priority to be sure! These messages always leave me baffled. One would never write a letter to an individual at a company and have the letter simply state:

---

432 Main Street
Atlanta, Georgia

Dear Ms. Richardson:
    I am interested in corresponding with your company. Would you please write back to me at the above address?

Sincerely,
Michael Hackett

Yet on a daily basis, people leave messages that merely announce who they are and ask for a callback. Clients tell me this is probably the greatest source of their voicemail frustration. Leaving vague messages also causes the receiver of the message to make assumptions about why the person is calling, and this may result in the call not being returned.

Running a close second as the cause of frustration are messages which ramble on and on, until the listener is so confused they have to rewind and pay close attention to minute details to make sense of the message. As E. W. Howe once said, "Half the time when men think they are talking business they are wasting time." Mr. Howe died in 1957 and did not live long enough to receive a long rambling voicemail!

Examples Of Informative And Effective Messages

"Steffi, this is Walter Brown. I'd like you to book me to travel from Atlanta to Phoenix on Wednesday the 8th with a flight to get me there by 1:00 p.m. I will need a rental car for three days, returning it Saturday morning, and that's when I'd like you to book my return flight. My hotel arrangements have been handled. I want to go ticketless. Please fax me an itinerary for my records. Thank you."

"Tom, this is Jodi Levy. I checked the specifications on the Encore 7300. It only requires a 110-volt power input, and we recommend a dedicated circuit. The footprint of the unit is 34" wide by 26" deep. The overall weight is 185 pounds, and it can be mounted on an elevated platform if you are short on space. If you have any further questions, please don't hesitate to call and leave me a voicemail anytime."

"Barbara, this is Brian Hogue. My logistics office verified that the thirty-six units were shipped this morning via two-day service. The tracking number is 102012151018. You can expect to receive them on Wednesday. Call me if I can be of any further assistance."

"Linda, this is Brenda with Beautytime. I received your message and I do have a supply of liquid foundation in 4 ounce bottles. I will leave two in a sack on your front porch later this afternoon, and will charge them to your credit card number that I have on record. Thanks for your business!"

"Mr. Clement, I appreciate your call and look forward to our meeting in the morning. Our office is located at 50 Cass Avenue; going north on the interstate, take exit 46, then turn right at the top of the ramp. Go to the second light, bear left on Faulkner and then the second right is Cass. See you at 7:30 a.m."

"Susan, the number for the open invoice we are tracking is 850626. It was for delivery of supplies ordered on your purchase order number 30390. If you need further information, please call me back and leave a voicemail if I am not available."

"Hello Ms. Stiles, Gary Grogan. I have two questions regarding the design changes to our new office plans. First, we'd like to increase the sitting area in the reception lobby by eliminating the storage closets, and we are wanting to add three generic work stations on the west side for our employees who telecommute and occasionally come to the office. Could you call me in the morning, between 8:00 and 8:30, to discuss incorporating these modifications?"

"Hi Barb, this is Paul. I will be arriving on AirTran Flight 626 at 4:40 p.m. Just drive through the lane on the lower level and look for me. I will be at the curb and have a bright yellow rolling bag."

"Bill, this is Vince. I need cost estimates and projected lead-time on the system we discussed, and need the info for a meeting at 1:00 p.m. Thursday. A voicemail will be fine."

"Good morning Bruce, this is Patricia Graham. I reviewed the issues with my manager and here is what we will do to alleviate your concerns: we will credit the delivery charges, schedule an additional training day for your staff on Monday the 24th, and dedicate a telephone support person to your company for the first thirty days. I hope this addresses your concerns. If you would like to discuss this further, please call me between 8:00 and 11:00 this morning."

"Hello, Jim. I wanted to set up a time when we could meet to review and discuss the bids for the addition to the Jacksonville plant. I want to spend a couple of hours together. I am open Wednesday at 9:00 or 2:00, and Friday at 1:00. Let me know which works best for you and I'll block my calendar."

"Hi Karen, it's Jim. 2:00 Wednesday works best for me. I'll see you then. You might want to invite Blair Johnson to join us by conference call from Jacksonville for his input. Goodbye."

"Dr. O'Donnell, this is Dr. Abbott. I am in family practice in Highland Village, and I am referring a patient to you who is a fifty-eight year old male complaining of neuromuscular tremors consistent with Parkinson's. The patient's name is Robert Jackson and he has no prior history relating to these symptoms, nor did I personally observe them during his visit today. My office has obtained prior approval from Mr. Jackson's PPO for you to evaluate him. I would appreciate hearing back from you after you have done a neurological work-up and arrived at a diagnosis. My telephone number is 555-8963. Thank you."

Message Planning Guidelines

1. Before you make a call or callback, jot down the relevant points of information you want to cover, questions you want to ask, etc.

2. Begin your message with a clear introduction of yourself, state the reason for the message, and ensure the information contained in the message will benefit them.

3. Speak slowly and clearly, especially if you are providing telephone numbers, facts, figures, or technical data. Most voicemail systems allow you ample time for messages, and warn you when a limited time remains.

4. If the system gives a warning that your time is almost up, do not speed up your rate of speech! State that you have additional information and will place a second call to complete the transfer of information.

5. Should a third party answer and offer to take a message, ask to be put into the person's voicemail box. Only you can deliver a complete and accurate message and present it with positive tone, enthusiasm, and energy. A third party will only capture the bare facts of the message, and may even make an error in relaying important information.

6. Leave the message while standing, not slouched in your chair. It will give greater depth and richness to the sound of your voice.

7. If the message is important, rehearse it by leaving it for yourself on your voicemail.

8. When the system offers the option, play back the message you left and be sure you feel comfortable with it. If not, erase it and start over. Remember that impressions go a long way!

9. Close the message with a request for action or follow-up by the other party that will keep the dialogue moving forward.

10. If the message is low priority and informational only, such as internal company or product data, inform the receiver of the message up front and suggest that they save the message and listen to it at an opportunity when they have more time.

11. If you have several topics to address, consider leaving separate messages for each topic. This will ensure that if the receiver should forward the message to another individual, that person will not have to listen to unrelated information.

12. Consider the possibility that your message may be forwarded to others; take care that it is structured to leave a favorable impression with *any* listener.

6
Chapter

IV

Fourth Commandment:
Project Positive Enthusiasm And Energy
In The Messages You Leave For Others

That which we are capable of feeling we are capable of saying.
— Miguel de Cervantes

Most of us feel annoyed by voicemail sometime during the workday, whether we experience a slight irritation or the deep burning sensation of *Voicemail Hell* in our stomachs. Few people smile when those they call are unavailable. Subconsciously, we develop anger or resentment towards the other person — *how dare they not be there, ready to take my call* — and our frustration builds.

Believe it or not, this frustration comes through loud and clear in the tone of your message, which is usually curt. The other party can clearly detect that you are unhappy about something. Perhaps they wonder, are you upset with them? Do you hate your job? Are you being deceitful in the information you are conveying to them?

The way we convey messages through our tone, enthusiasm, and emphasis on words clearly has an impact on how that message is received.

The following statement, with different words emphasized in bold, illustrates this notion. Read each one aloud, emphasizing the word in bold print.

You are going to sit with us? That's great.

You are going to sit with **us**? That's great.

You are going to **sit** with us? That's great.

You are going to sit with us? That's **great**.

The last example makes the person feel welcome, while the first three all emote some kind of sarcasm. Remember, it is not always the words you speak but how you speak them; the tone and inflection shape your message a great deal.

To support this, consider the results of a study by Dr. Albert Mehrabian of UCLA. According to Dr. Mehrabian's research, people form impressions of us according to the following proportions: — our body language 55% — our tone of voice 38% — the words we speak 7%.

When we communicate with others by telephone, however,

they are not in a position to read our body language, therefore tone becomes much more important as it accounts for 87% and words for 13% of their impression of us.

You can see that the tone of your message, if it projects positive energy, can go a long way in creating a more favorable impression of you on the part of the listener.

While it is difficult on paper to accurately depict the difference between curt and positive energy messages, a comparison of the following should illustrate the difference.

"Brian, this is Jan Harrison returning your call again. Please get back to me when you have a chance. Thank you."

"Good morning, Brian! This is Jan Harrison. I'm sorry to miss your call, and I'd like to know how I can help you. If you miss me when you call back, please leave me a message detailing your concerns so I can begin to take action. Thanks, and I look forward to helping you resolve these issues."

Now, which of these messages makes Brian feel like Jan wants to help him, and which makes him feel like he is putting her out?

Guidelines For Projecting Positive Energy In Messages

1. When you hear the voicemail kick in, do not get upset. Take a deep breath, and do not think of it as a missed opportunity for live conversation, but rather as an opportunity to deliver important information without interruption.

2. Smile as you speak. Guaranteed, the smile will project itself in the tone of your voice, and your enthusiasm will come through loud and clear. Place a small mirror on your desk to facilitate this. See yourself brooding and you will quickly change your facial expression to a more pleasant one.

3. Stand and deliver. Standing, as opposed to sitting in your chair, will allow you to enrich the sound of your voice.

4. A small piece of well-chosen, lighthearted humor can add enthusiasm to the message.

5. If one is available, have a picture of the person you are leaving the message for handy. Speak as though you and the other person were face to face. (Pictures from company gatherings, customer events, and corporate newsletters are generally readily accessible in most people's work area.)

7
Chapter

Fifth Commandment:
Train Others To Communicate Effectively
With You Via Voicemail In Order To Eliminate
"Telephone Tag"

Getting results through people is a skill that cannot be learned in the classroom.

— J. Paul Getty

I wish I had a nickel for every time someone has received a voicemail message with the cute remark, "Tag, you're it!" I could retire. Clients have said to me, "Gosh, I've gotten so good and efficient at using my voicemail effectively, but it's the other people who regularly drive me crazy. They leave pointless messages that still require me to track them down to get to the bottom of issues!"

What this problem requires is peer or customer education. I remember when I first started asking in my greeting for callers to *"please leave me questions and details in your message so that I can get back to you with the information and answers you're looking for,"* I would get many hang-ups. But to my pleasant surprise, the hang-ups would be followed just a few minutes later by a second call, one that was planned and somewhat scripted according to my request; there were questions and details and I was able to move the dialogue forward in my follow-up callbacks. I realized that as I communicated to others how best to communicate with me telephonically, our voicemail dialogues became extremely productive.

The kind of message that is frustrating typically sounds like this: *"Paul, this is Theresa at Ozark Corporation. Could you please give me a call when you have a chance? I have the details on the sales training class you'll be teaching for us next week and I want to give them to you."*

When I receive such a message, I return the call after hours and inform the other party that in order to save them time trying to connect live, why not just call my voicemail and leave all the detailed information. I ask them to leave me the name of the hotel, my confirmation for overnight accommodations, start time, number of attendees, name of onsite contact, and other pertinent information. I wait patiently and am usually rewarded with an informative message, generally no longer than a minute. This process certainly adds to my time management efforts.

Messages That Establish Effective Two Way Communications

"Noel, I know how hectic your schedule can be. I don't need to hear from you live. You are welcome to leave me voicemails with any information you have, rather than trying to catch me at my desk."

"Mr. DeVane, I understand that you have questions regarding our past service on your system, and I am sorry we keep missing one another. If you wouldn't mind leaving the details and your questions on my voicemail, I can research the answer and get back to you with the correct information."

"Ms. Perez, this is Byron Siske. In an effort to better serve all my customers, I encourage the use of voicemail to communicate with me. You can leave a detailed message for me anytime, 7 days a week. That way, I can get back to you with the information you need."

"Hello Mr. Fuqua, this is Theresa Abernathy. I realize that your responsibilities take you away from your office for most of the day. I'd like to suggest that we use voicemail to correspond with one another with respect to any questions or details which may arise as we move forward with this project. This way, we will remain on schedule.

"Since my voicemail is active 24 hours, feel free to leave me messages any time a question or issue arises, and I'll promptly follow up with a response."

Guidelines For Effective Two Way Voicemail Communication

1. Ensure that your outgoing greeting contains a request for information or a "call for action".

2. Always leave well-organized, informative messages free of clutter and confusion.

3. Tell the recipient that you welcome their response via voicemail, that ear-to-ear conversation is not necessarily required.

4. Inform the individual at the start of the relationship how you like to correspond telephonically.

5. Discuss whether voicemail is agreeable for the routine exchange of information.

6. Inquire as to their voicemail's 24-hour capabilities; if yours is always accessible, inform them so.

7. Give them a personal copy of *Escape From Voicemail Hell!*

8
Chapter

VI

Sixth Commandment:
Use Internal Voicemail And Group Broadcasts
To Maximize Communications

The telephone is a good way to talk to people without having to buy them a drink.

— Fran Lebowitz

To paraphrase Ms. Lebowitz' quote, voicemail is a good way to communicate with people without having to talk to them directly!

When you first signed on to your voicemail, you were probably so confused by the process of setting up your password and greetings that you never read below the first few options on that complex instructional flowchart. Internal voicemail can be extremely effective in helping you manage your time and enhance your internal work relationships.

Better information management with colleagues can be achieved by utilizing the internal send option on voicemail. Clients tell me that a good deal of telephone tag takes place internally. They also describe times when they have a simple question that requires a brief answer, yet the person they call draws them into a lengthy conversation. This consumes valuable time and distracts them from the task at hand.

Rather than placing a call to the person's telephone, why not utilize the internal send option on your voicemail? This way, the person receives a clear and concise message that outlines the required response. They can gather the information and respond back to you directly via voicemail generally by just pressing one button on their telephone. This allows both of you to manage time and calls effectively. Low priority internal messages can also be deferred until a quieter time for reply.

Look in the very bottom of your top left desk drawer for that voicemail system flowchart, like the one illustrated back on page 5. It's probably buried underneath the pens, pencils, the individual packages of crackers, and artificial sweetener . . . ah, there it is! Examine it closely; somewhere in that maze of squares, rectangles, and arrows, there is probably a feature called "group broadcast", "group list" or a similar title.

The group broadcast function gives you the ability to multiply your message reach, thereby saving time and increasing awareness and team spirit in your work group. It is also much quicker than typing and sending e-mail, and your message is delivered with the positive tone and enthusiasm of your voice!

Once you have set up your group lists, a message can go out to large groups of people in seconds simply by pressing a couple of buttons. (Always take care to ensure, however, that the message goes out to the right group!) Did Peter make a sale or turn around a dissatisfied customer? Put out a group broadcast! Team meeting on Thursday at 2:30 p.m., or department softball tryouts Monday evening? Broadcast the information.

Did you suddenly come upon competitive information or new sales information about your product? Tell your whole team with the push of one button. Want to motivate your team? Give them a daily pep talk via a broadcast message! One client in the telesales business told me that every morning he puts out a message congratulating the previous day's sales leaders and a positive and motivating "thought for the day". Talk about building team spirit!

The following examples may be used as a guide for scripting your own internal messages and group broadcasts.

Internal

"Charlie, this is Mike in accounts payable. I am processing invoices from Seaside Paper Company and wanted to verify that we did receive the shipment of 150 cartons of copy paper on October 20. I do not have the signed paperwork showing it was

received. Could you get back to me with confirmation and if the receiving documents are still in your office, please send them up to me? Thanks."

"Mike, this is Charlie getting back to you on the Seaside Paper invoice. Yes, we did receive the 150 cartons of copy paper, and I do have the paperwork still here. I will send it up to you via internal mail. Sorry for the delay."

Group Broadcasts

"Good afternoon folks, I'd like to take a couple of minutes to share a success story with you about Elizabeth and how she overcame the challenge we were facing in the design of the new system architecture. After re-examining the response configuration . . . Elizabeth's efforts in this regard will save us weeks of downtime, and I hope you all will let her know how we appreciate her effort and results."

"Just a quick reminder to all that we have a staff meeting at 8:30 tomorrow morning. Please come prepared to give a ten-minute overview presentation of the current status of your project. Christina Godin, our new VP of operations, will be joining us to give us the chance to get acquainted with each other and to bring her up to speed on where we are with the various projects. Please be prompt and well prepared for questions."

"Congratulations to Mark Thomson for wrapping up a $32,000 sale of a System 9600 to Pace Corporation. Mark has been working on this opportunity for over five months. With the help of Todd and Judy he was able to bring it in. Great work, team, now let's keep it going!"

"The warehouse crew has challenged us in customer service to a beach volleyball match at the company outing later this month. Anyone wanting to join our team and help us bury the warehouse crew in the sand, please put your name on the sign-up sheet outside my cubicle."

"Good morning! The forecast for the day is sunny and clear so I hope it is a good day for all of you, and I hope you take advantage of the great weather to walk the campus or have lunch outside by the duck pond. Something to think about today - a quote from Aristotle, 'we are what we repeatedly do, excellence, then, is not an act, but a habit'."

"Hi! This is Tom Archer. As CEO of the company, I want each and every one of you to know how proud I am of your contribution to this year's United Way Campaign. Your dedication to this very worthwhile cause shows that all of our associates are committed to helping our fellow human beings who may be in need, and it makes me very proud to be associated with such a fine group of people. Thank you and God bless you!"

"Hello, gang. I received the following voicemail from a customer recognizing Tamsin for a job well done. I am forwarding it on so that each of you can hear it and perhaps gain some perspective from it" . . . ["Hello, Ms. Wallace, I am calling to commend one of your staff . . ."]

"Good afternoon, team. I want to share with each of you a bit of competitive data we just received about Siler Engineering and their new hydraulic pumps that compete with our Astro series. It seems that the pumps do not quite live up to their advance billing, based on feedback that we received from a customer who had made a switch. What the customer tells us is that the impellers in the pumps are made of an alloy that does not have the strength of our solid steel impellers."

Guidelines For Effective Internal Voicemail

1. Always prioritize internal calls. Is the information critical at the present time, or can I wait to respond to the message?

2. Send low priority messages via internal voicemail rather than live calls.

3. Return internal calls via voicemail, and defer low priority callbacks to off-hours.

4. Set up group call lists within your voicemail; dedicate group lists to team members, colleagues, and different management levels.

5. Use the group broadcasts regularly, but ensure that the information passed along is timely, informative, and presented in a clear and concise manner.

6. Always forward good news and commendations throughout your group broadcast network since hearing a complimentary message in the original caller's voice is much more motivating than a second hand account.

9
Chapter

Seventh Commandment:
Make Use Of The 24x7 Capabilities Of
Voicemail

Don't forget until too late that the business of life is not business,
but living.

— B. C. Forbes

This is the commandment that always presents an interesting
contrast. Clients' first reaction is usually a harsh rebuke. "What? I
work myself to the bone all day long, and you expect me to work
at night and on weekends?"

However, once they give the concept consideration and agree
to try it out, they call back with rave reviews! *"I was able to get out
of the office at 5:30 even though I had six voicemails to respond
to. After the kids were in bed, I played back and responded to*

those six messages from home in less than 15 minutes. Had I connected with only one of those individuals live before I left the office, it might have turned into a twenty minute conversation."

Another client stated, *"On Sunday evening I leave voicemails for everyone on my staff, telling them my expectations for the coming week and outlining any major issues they need to address. What used to take two hours of face-to-face rounds on Monday morning now takes less than 30 minutes on Sunday evening, or early Monday morning before I leave home."*

I am not suggesting that you bring all of your work home and avoid live contact with people. Remember that voicemail is simply a tool to efficiently transfer information in a timely manner. Certainly there are times when live face-to-face or ear-to-ear contact is necessary. Yet, if you think about it and start reflecting on and prioritizing your messages, you will realize that not all callbacks are time critical or require live interaction. If they are merely requests for information or clarification, they do not require a live response. They just require an *informative* response that keeps the dialogue moving forward.

One client's feedback described making internal callbacks from home in the evening and seeing the time it saved him. He even began making external callbacks from home in spite of having to pay long distance charges. *"My long distance calls are only ten cents per minute. For less than a dollar, I can usually make half a dozen callbacks since the messages I leave are always brief* and to the point. It beats tying up an hour or more calling people live during the day, and it gets me home earlier." Certainly this is testimony to the power of this principle.

Another benefit of making callbacks during late weeknight hours, or on weekends, is that it will make quite an impression on the person you are calling. *"You were working on my problem at 10:30 at night? Wow, I am quite impressed!"* (Learn more about this in Chapter 10.)

One client, who is in computer sales, works her voicemail and does her callbacks during the 10:00 p.m. news, and sometimes in the morning after she finishes her 6:00 a.m. workout. *"It's peaceful and quiet in my home office then, and I can knock out many calls quickly."* She sees it as an efficient way to communicate information, and uses her live time with customers to address key issues, present new products and grow her relationship with the customer.

This client has also come to realize that by utilizing voicemail during off-hours, she has been able to achieve a 30% increase in sales of add-ons and upgrades to her existing customer base!

Guidelines For Using 24x7 Voicemail Capabilities

1. Evaluate your incoming calls as to required response, live or via voicemail.

2. Prioritize the callbacks according to their value. (Second Commandment)

3. Determine which of your business contacts have 24-hour voicemail. (Home-based individuals may have it, but the telephone may ring in their bedroom!)

4. Ensure that others are aware that for efficiency sake, you sometimes return calls during evening and weekend hours. (Fifth Commandment)

5. Select a quiet time, when family members are asleep or out, or a quiet place with no distractions, and make your callbacks.

6. Enjoy the extra time gained by leaving the office promptly at the end of the day; share that time with your family or other loved ones, or pursue an interest or hobby.

10
Chapter

Eighth Commandment: "Wow" Others Via Voicemail And Build Stronger Relationships

Measure a person by the stretch of his imagination.

— Robert Schuller

A great deal has been written about how to differentiate yourself from the pack, how to impress customers, colleagues, peers, etc. Let me suggest doing it via voicemail!

Remember, people love surprises and things or actions that are unique. In today's business world, suppliers and vendors have become clones of one another. They employ tried-and-true techniques to follow up with customers or others to acknowledge a purchase or other significant milestone in the relationship.

Cards, notes, flowers, and other marketing paraphernalia have been the usual method of showing a customer appreciation or acknowledging a colleague's contribution. Here again, voicemail, and its ability to allow you to deliver your message with positive tone and enthusiasm can set you apart.

Just think how special a person might feel dialing into work on the Monday after Thanksgiving and hearing a voicemail message that you left on Thanksgiving Day telling them how much you appreciate and are thankful for the business relationship you have with them? Would that differentiate you from the pack?

What kind of impression would you make on a colleague who assisted you with a major project if you left them a message one evening acknowledging their effort and contribution? The message may linger for days so they can listen to it time and again. Remember the words of Mother Teresa who said, "Kind words can be short and easy to speak, but their echoes are truly timeless."

Would a business acquaintance be flattered and impressed if you left them an off-hours message congratulating them for an accomplishment by one of their local sports teams?

Would your reputation as a caring, concerned individual be enhanced if a colleague or customer in a locale that suffered a natural disaster received a voicemail message of support from you?

How about a person dialing into their voicemail upon arriving at work on their birthday, and hearing your 12:02 a.m. message saying, "I wanted to be the first to wish you a Happy Birthday!"

What kind of an impression would that make?

One client took this suggestion to heart and called his fifteen top clients on New Years Day to leave them voicemails. The messages stated that he appreciated their business, looked forward to working with them in the coming year, and hoped they were having a good holiday. On his first day back to work, he received calls of acknowledgement from all fifteen, including two who told him that they were going to make him their sole supplier for being so considerate!

In order to "Wow" people via voicemail, you are only limited by your creativity and imagination, and remember that the 24x7, 365-day capabilities of voicemail provide you with many opportunities.

Examples Of Some "Wow" Messages

—Retrieved on January 3—

"Hi Mr. Kirby, this is Mark Parks with Century Resource Group. I'm leaving this message on New Year's morning. I hope that you and your family are enjoying the holidays as much as my family and I are. I wanted to wish you a Happy New Year, and let you know how much I appreciate the business we have done together the past several years. I'm looking forward to a great year working together! I'll talk with you soon."

—Left late one October evening—

"Hello Gayle, this is Theresa Randall calling. I just watched the New York Yankees win the World Series. I know what a big fan you are, so I wanted to pass along my congratulations. I'll talk with you Monday morning."

—Left at 6:15 a.m.—

"Vincent, good morning this is Kurt with Osherman Supply. I just read in the morning paper that a new company called Millennium Fabricating Corporation is planning to build a facility in this area and hire several hundred employees. I know this could be a potential opportunity for you so I wanted to bring it to your attention. It's on page 3 in the metro section."

—Left at 9:45 p.m.—

"Good evening, Mr. Kline. This is Sara Hoffman with Monterrey Products. I'm calling to again thank you for placing an order with our company, and to let you know that I confirmed with our dispatch department that the product was delivered in the quantities you designated to the three separate locations. We appreciate your business and look forward to serving you again in the near future."

"Hello, Lydia. This is Marvin Alworth in the Denver office. I've been following the news of the tornadoes in your area and hope *that everything is okay with you and your family, as well as the rest of the staff there in the office. You are all in our prayers. Please let us know here in the Denver office if we can help you in any way. Good luck and stay safe."*

"Hi, Mr. & Mrs. Taylor. This is Marcia Simpson. I'm standing here in my office window watching you drive off with your new vehicle. I want to thank you again for your business, and please be sure and call me if I can be of any assistance at all."

"Hey Kevin, this is Warren. I want you to know how much I appreciate your staying late today to help me review the spec sheets on the new equipment. Your expertise saved me a few hours of studying and trying to make sense of them. It really makes life easier working with team players like you. Thanks, Bud!"

Get the "message"? You are only limited by your willingness to employ your creativity and enthusiasm in order to "Wow" the people you deal with and enhance your relationships with them. Go ahead, stretch your imagination!

Guidelines For Wowing Others Via Voicemail

1. Learn as much as you can about the business and personal interests of the people in your business environment.

2. Pay attention to news and business events and consider how they impact those in your circle of business contacts.

3. Ask yourself, "what would make me feel appreciated and special," and follow the Golden Rule.

4. Take every opportunity to differentiate yourself in the eyes of your clients, colleagues, and superiors.

5. Remember that a fifteen-second voicemail message can reap benefits long after it has been erased.

11
Chapter

Ninth Commandment:
Harness The Power Of Voicemail
To Prospect For New Customers

If you don't sell, it's not the product that's wrong, it's you.

— Estee Lauder

My years of working with salespeople have taught me one major fact: salespeople dread making cold calls, especially telephone cold calls to secure a first appointment. Prospects cut them off in the beginning of the call and barely allow the salesperson to deliver their "well-rehearsed pitch" that they feel will get them face time with the prospect.

We really cannot blame the prospects. They are usually busy and involved in multiple tasks when a telephone solicitor calls, probably the 5th or 6th of the day who wants them to drop everything and listen to their pitch!

The call usually starts like this: *"Hello, Ms. Austin, this is Marc Shelby with Vega Corporation. How are you today? Am I catching you at a bad time? I'm calling to tell you a bit about my product and make an appointment to come and talk to you so I can show you how my product can benefit your company by reducing your production costs and saving you money."*

The salesperson is focused on themselves and their product, not on the prospect or their situation. "How are you today" and "am I catching you at bad time" are both patronizing. The salesperson hasn't made a business case for the prospect to want to listen. When the salesperson tells the prospect their product will benefit the prospect's company by cutting production costs and saving them money, it sounds insincere. How can a promise be made when little is known about the prospect's situation? Meanwhile, the prospect's attention is focused on other matters, and the call ends quickly.

I once asked a group of telesales people when and how they engage in prospecting for new customers. The general response was "when I have a few spare minutes between scheduled calls." They treat cold calls almost as an afterthought or time-filler! They do little or no preparation, read a quick introductory script to the prospect and then wonder why the person is not interested in their pitch. What a pitiful first impression, and we don't get a second chance to make a first impression!

I recommend that if you seriously want to prospect new customers by telephone, you prospect in the evening by voicemail. The prospects will not be there to hang up on you, and your message will be one of the first things they hear at the start of the next business day before they get involved in multiple tasks and issues. Pulling together all that you've read thus far in previous chapters with additional strategies, you can become a proficient prospector.

Your voicemail message has to be powerful enough to capture their attention and make them want to call you back, or take your follow-up call. It should not sound like a standard telephone call with you introducing yourself and company at the beginning but more like a personalized radio advertisement, with a "grabber" up front related to their business. Do not pitch your product; tell them that you would like to offer them more details about the grabber you shared and that your company offers solutions to address the specific situation mentioned.

The grabber should be a business fact or statistic that affects their business, and if possible one about which they may not yet be aware.

A client who sells industrial supplies was encountering the usual frustrations playing the "dial-and-smile" game seeking new prospects. He was receiving a myriad of negative responses to his calls and was becoming discouraged, so he decided to try this technique.

He selected several prospective customers in one industry and researched industry issues on the internet. He then structured a high-impact message and left it on the prospects' voicemail in the evening. The result was that he was successful in gaining live follow-up calls with two of the five prospects. He is a true believer in this technique.

The technique can work for you too. Remember though, it's not about just leaving a message in the evening. It's about structuring a message that addresses their situation or challenges, not one that boasts about you and your company.

The following examples illustrate how a well-structured voicemail message can grab the listener's attention:

"Ms. Bishop, good morning. Are you aware of the recent changes in state law that affect local telephone service? Did you know that companies are now free to choose from a number of communications providers and that they can customize their service architecture, and in the long-term reduce their telephony expenses?

"My name is Louise Smotherman and I represent Waterside Communications, a new competitive local exchange carrier. I'd like to provide you with more information about your new telecommunications options, and will call you today, Thursday, at 10:30 a.m. to schedule a time when we might meet to discuss those new options. If this time is not convenient for you, please call me or leave me a voicemail message at [411] 555-1018 to let me know a better time that we could speak to arrange an appointment."

"Mr. Keller, a recent study by an industry group shows that on average companies with 1,000 employees lose 3,575 employee work days annually due to family illnesses where the parents have to stay home to care for a child who can't be taken to school or daycare. Is this problem discreetly affecting productivity within Lawson Technologies?

"My name is Lester Shaeffer, and I represent Childcare Calamities, Inc. We specialize in assisting companies like Lawson to deal with this very costly problem. I would like to call you today to perhaps arrange a time when we could meet and I could give you further data from this study as well as some information about CCI. I will call you in the early afternoon. My direct number for your information is area code [769] 555-5243."

Note that both of these messages grab the person's attention with a business fact related to their situation, then set up the callers as sources of additional information and possible solution to the problem. They show the prospects that the callers are interested in more than just talking about themselves and their companies, they are in tune with the prospects' business issues.

Effective Guidelines For Voicemail Prospecting

1. Dedicate a specific timeframe in the evening, perhaps after 7:30 p.m., to make calls.

2. Select a quiet location to call from, preferably your office or in your study at home.

3. If you plan to call from your office, take a break at the end of the day; go out and have a good dinner and relax as you prepare to return and make your calls.

4. Research your prospects and their business; seek out facts or statistics that are relevant and that can lead to your company's products or services as a solution.

5. Plan your message according to the guidelines in Chapter 5, the Third Commandment. Structure your message so that it opens with the person's name, your grabber, a brief mention of your company, a number for them to call you, and also a projected time that you will call them again live.

6. Be aware of the person's level within their organization, and structure the message accordingly.

7. Follow-up with a live call to the prospect the next day!

8. Be respectful of the prospect's time and workload when you make your follow-up call. If appropriate, make an appointment for a live telephone conversation at a future date.

12
Chapter

Tenth Commandment:
Manage And Enhance Your Personal Life
With Voicemail

Leisure is the time for doing something useful.

— Benjamin Franklin

My wife and I are both working professionals with busy schedules, children, church and community activities. Our time together is valuable, and we hate to spend it dealing with housekeeping matters. We probably communicate via voicemail with each other more often than with any other party.

Since we are both home-based, we are on the same voicemail system through our local telephone company, GTE. This makes it easy to send each other internal messages whenever we dial-in to check our messages.

We both spend time at client sites and each of us travels on occasion. Whenever a travel engagement arises for one of us, we immediately voicemail the other to block schedules so we are not out of town at the same time. We also use voicemail to pass incidental matters along to each other. What time is soccer registration? Who has carpool duty for the children's religious education classes this week? How much was the estimate for the house painting? Which movie is playing where and at what time? How did the meeting go with that client? While these and other messages may seem incidental, they add up to lots of information to exchange if we save them all for evening. We would rather handle all these matters and transactions between ourselves during our workday downtime so that our evening family time is devoted to our quality of life.

Consider leaving reminder messages for yourself on your own voicemail. Rather than writing a note to yourself only to risk misplacing it, you can leave yourself a reminder message at work that will be there for you at the start of the next workday.

Susan and I also utilize voicemail to keep our romantic sparks alive and well. It only takes a few seconds to leave a sweet, romantic, or empathetic message that can go a long way in enhancing your relationship. Many times I check my voicemail while traveling and am delighted to receive a romantic message left for me. I save it and listen to it over and over again until I return home, or until a subsequent message is left for me!

When either of us travels, we regularly send voicemail messages to our children. Justin, the fifteen-year-old, listens once then deletes; eight-year-old Jared will replay it five times and then dial-in thirty minutes later and rewind it another five times!

It is also a treat to receive a message from one of the children. That is always good as a pick-me-up during a challenging day.

I cannot begin to tell you how many clients have reacted with "Oh, Gosh! My spouse has voicemail. I have never even thought to leave them a romantic message to brighten their day!"

Believe us, speaking for Susan and myself, your limitless creativity will get you mileage.

Examples Of Personal Messages

"Hi Amy, hope your day's going well. I just made plans to be in Chicago April 3-5, so can you block that on your calendar? See you tonight, love you."

"Hey it's me. I wanted to tell you again how much fun this weekend was. It was great getting away by ourselves and having such a romantic time. That inn was so cozy. I can't wait to go back sometime again, just the two of us. Have a great day, sweetie!"

"Hi Sean, I'm scheduled to pick the kids up at daycare and get them to baseball practice but my meeting is going to run late. Can

you leave me a message if you can get there by 5:30? Please let me know by 3:00 so I can call Beth to pinch hit. Bye."

"Hi Peter, the contractor came by this morning and gave me an estimate for replacing the landscaping around the pool. I showed him our design, and he said he could do all the plants, timbers, and fill for $850. He is booked about ten days out, so if you agree with me that this is a good price, call me and let me know so that I can book him. Hope your day is going well. Kristin says, Hi, Dada!"

"Hello, honey. I hope you're feeling better and that you were able to address the issues with Mr. Walker. I know these things have been on your mind and I want you to know that I support you and understand the frustration you are feeling. Remember, I'm always here for you."

"Reconfigure the service proposal for Greenhouse Supply utilizing their in-house tech support people for level one issues." (Message to self)

"Hey dude, congratulations! I heard about the awesome goal you scored at soccer. I'm sorry I wasn't there to see it. I'm looking forward to seeing you play your next three games. I'll see you this weekend. Don't forget to put out the trash on Friday!"

"Hey, how's my big guy doing? Daddy is in San Francisco and today I drove across the Golden Gate Bridge and got to see some giant redwood trees. I sent you a postcard with a picture of the bridge on it. You be a good boy for Mom, and I'll be back home on Wednesday. You be good in school, and be sure to help Mom! Here's a big smooch coming your way!"

"Hi Mommy, it's me, Adam. I wanted to tell you that I got a special award at school today for being the first to finish my project. I can't tell you what my project is, 'cuz it's a surprise for you and Daddy for Christmas! I love you and I miss you, Mommy! Bring me back something!"

"Hey Dad, I got some information today in school about a summer study program in Washington, DC. It's a two-week program where you get to learn about the government; you meet with members of Congress and foreign ambassadors, and some of the reporters who cover the White House.

"I'd like to talk it over tonight during dinner. Oh, by the way, it costs $2000. See you later."

13
Chapter

Putting It All Together

Remember, nothing that's good works by itself, just to please you.
You've got to make the damn thing work.

— Thomas Edison

So there you have them, the *Ten Commandments of Voicemail.*
Do they all fit your specific work situations? Can you make all or
some of them work for you? These questions can only be answered
over time.

As it was suggested at the start of this book, let the ideas and
material and the principles sink in a bit. Come back and read it
again in a few days and make note of the pages you feel would be
most helpful to you.

Begin by making just one change in how you communicate via voicemail.

- Develop a new and improved greeting that will impress your callers and prompt them to leave a call for action.

- Leave someone a message late in the evening or early in the morning that relays critical information to him or her. Then sit back and wait for their callback telling you how impressed they are that you were thinking of their issues at that time of day.

- Begin the workday with a broadcast message to your peers or subordinates that acknowledges their day-to-day contributions. They may replay it time after time as a boost to their spirit.

- After a loved one heads off to work and before they arrive at their office, leave them a message telling them how special they are to you.

- When you leave someone a message, state the purpose of the message up front, make the message brief yet informative, and tell the other party what you need from them to keep the dialogue moving forward.

- Leave the office on time, and manage your callbacks later in the day when activities settle down in your personal life.

- Send a voicemail to a prospective new client that whets their appetite for more information — that only you can provide.

- Send that important message to your own voicemail before you send it to your boss or customer and see how it sounds.

- Drop a message of encouragement at home to that teenager

coming home after a day that included a calculus exam and basketball practice.

• Make sure that every message you leave is upbeat and enthusiastic; do not drag someone down, but rather pull the person up onto a higher plane.

Whichever techniques you choose to employ, you will have to practice them faithfully for twenty-one days. It's been often said that is how long it takes to establish a habit or pattern.

Improving your telephone communications, managing your calls and callbacks, and becoming more productive are not merely the result of reading a book. Becoming a more productive person requires a commitment on your part, because it is not just about doing certain things, but about having a state of mind, or attitude, that says, *"I want to be more productive!"*

By adopting such an attitude and employing the *Ten Commandments of Voicemail*, peers, colleagues, clients, and loved ones will view you in a different light. They will see you as a person who knows how to get things done effectively, and who can keep a dialogue moving forward while maximizing their communication contacts. You too will feel better and more productive, the frustration and burning sensation will disappear, and you will never again burn in *Voicemail Hell*.

Afterword

While in the process of writing and developing this book in early 1999, I was involved in a project with the PTA at my eight-year-old son's school, Christa McAuliffe Elementary, in Highland Village, Texas.

The school was planning to dedicate a permanent exhibit in its library detailing the childhood and elementary school years of its namesake. Christa McAuliffe was America's Teacher-In-Space who perished in the 1986 Challenger Space Shuttle disaster. Christa's mother, Grace Corrigan, was invited from the Boston area to attend the dedication and to speak to teachers from several area schools at two assemblies.

I was responsible for securing a sponsor to underwrite the cost of Mrs. Corrigan's transportation and for receptions at the two schools where she would speak. I sought Ms. Connie Yates, Director of Public Relations for Tom Thumb Supermarkets in the Dallas-Fort Worth area. My initial inquiry to her was left on her voicemail late one evening, and it advised her that she could call my voicemail anytime 24 hours, since I was often away from my office. As it turned out, she had an equally hectic schedule.

Over a fourteen-week period, Ms. Yates and I exchanged countless voicemails with inquiries, responses, and lots of other pertinent information. Working in tandem and both relying on voicemail to manage our time and busy schedules, we left each other messages during the daytime, evening, and even on weekends.

Each message kept the dialogue moving forward. Both Mrs. Corrigan's visit and the dedication were very successful with Tom Thumb as the principle sponsor underwriting the event. Ironically, Ms. Yates and I both had to travel out of town on business during portions of Mrs. Corrigan's visit.

I received a nice note from Ms. Yates afterwards acknowledging how well we worked together via voicemail. I'm still looking forward to meeting her face-to-face or even speaking with her live on the telephone!

It's amazing how productive two people can be and what may be accomplished by effectively using voicemail.

About The Author

Paul LeBon, husband, father, and author, is a productivity consultant and President of lebon directions.

Paul has over twenty years experience in the field of sales and sales management, and for the past decade has been engaged in corporate education and consulting with many firms, including several Fortune 500 companies.

Additionally, he has served as an adjunct faculty member at Franklin Pierce College and Texas Woman's University. He holds a Bachelors' Degree in Marketing from Franklin Pierce College, and an MBA from Kennesaw State University. His professional affiliations include the American Society For Training and Development.

When not consulting in the corporate arena, he engages in numerous volunteer efforts, including building housing for the needy, adult leader in his church's youth ministry, and tutoring and mentoring at-risk children at a Dallas inner-city elementary school.

lebon directions is a productivity consulting firm founded in 1992 and headquartered in the Dallas/Fort Worth area, concentrating in the areas of sales skill enhancement, sales management development, and of course, increased, productivity through effective use of voicemail. For information about workshops and speaking engagements, please phone [972] 966-8135.

The Ten Commandments Of Voicemail

I.   Present a clear and concise greeting on your voicemail, including a "call for action".

II.   Utilize voicemail to effectively manage and prioritize your callbacks and your time.

III.   Plan in advance of calling so that your message is informative and effective.

IV.   Project positive enthusiasm and energy in the messages that you leave for others.

V.   Train others to communicate effectively with you via voicemail in order to eliminate "telephone tag".

VI.   Use internal voicemail and group broadcasts to maximize communications.

VII.   Make use of the 24x7 capabilities of voicemail.

VIII. "Wow" others via voicemail and build stronger relationships.

IX.   Harness the power of voicemail to prospect for new customers.

X.   Manage and enhance your personal life with voicemail.

Your Purchase Helps Support These Worthwhile Organizations

There are two organizations that are very dear to this author. They are Christian Community Action in Lewisville, Texas, and The Children's Rights Council in Washington, DC.

CCA provides food, clothing, shelter, financial assistance, affordable housing, and job and life skills training to needy families in the North Texas area.

CRC is an advocacy group for children of divorced, separated and unwed parents. Its aim is to ensure a child's right to the continuous love and support of both parents, regardless of the parents' marital status. CRC believes the best parent is both parents.

ParLeau Publishing will donate a portion of the proceeds from the sale of *Escape From Voicemail Hell* to each of these organizations. Donations to the CRC will be made in memory of the late Horace "Sonny" Burmeister, a tireless advocate for children.

For information on either of these organizations, they may be contacted at the following:

Christian Community Action
200 South Mill Street
Lewisville, TX 75057
[972] 436-HELP

Children's Rights Council
300 I Street, N.E., Suite 401
Washington, DC 20002-4389
[202] 547-6227

Volume Purchase Discounts

This book is an ideal productivity tool for all employees who spend any time communicating with others via the telephone. It can not only boost their productivity, it can help reduce telephone costs as well. Substantial discounts are available for quantity purchases of *Escape From Voicemail Hell.*

If a sufficient quantity is ordered, the book may be customized with your company's logo on the cover, and a member of your company's senior staff might write a custom Preface to be included in the text of the book. What a wonderful gift idea to reward your employees and enhance their skills!

For further information on investing in a tool which will boost your employees' productivity and help reduce telephone calling costs, please call ParLeau Publishing at [972] 966-8135 and ask to speak with our director of corporate sales.

Complimentary Productivity Aids

We sincerely appreciate your purchasing this copy of *Escape From Voicemail Hell*. We hope it helps you to be more effective and productive.

As a token of our gratitude, we would like to send you a free magnetic copy of the *Ten Commandments of Voicemail* to post on a file cabinet, storage bin, or your refrigerator. We will also include a peel-and-stick version for your telephone or day-planner, and a bookmark.

To receive these complimentary productivity aids, send a stamped ($.55) self-addressed #10 envelope, along with a business card or the following information printed on a 3x5 card:

Name

Title

Company

Mailing Address

Telephone Number (including extension)

E-mail Address

Our mailing address may be found on page 75. Your productivity aids will be mailed to you promptly.

Want To Be A Book Reviewer?

Now that you have completed reading *Escape From Voicemail Hell*, we would appreciate your feedback and input for possible future revisions. We have set up an option on our website that allows you to write and submit your own book review.

You can also tell us how *Escape From Voicemail Hell* has helped you. Want to share your new improved greeting with us? Tell us about a Wow message, a 24x7 voicemail story, or a successful cold call made via voicemail?

We welcome all of your comments, feedback, and suggestions on our website:

www.voicemailhell.com

Help A Colleague, Friend, Or Loved One Boost Their Productivity

Check your favorite bookstore or order here

Yes, I want_____copies of *Escape from Voicemail Hell* at $11.95 each, plus $3.00 shipping per book. (Texas residents add $.90 sales tax per book.) Canadian orders must be accompanied by a postal money order in U.S. funds. Allow 15 days for delivery.

My check or money order for $_____is enclosed.

Please charge my VISA_____MC_____

Name _____

Organization _____

Address _____

City/State/Zip _____

Phone_____E-mail _____

Card # _____Exp. Date _____

Signature _____

Print names of individual(s) for whom the book(s) should be personalized:

_____

_____

If paying by check, please make your check payable and return to:

ParLeau Publishing, P. O. Box 292845

Lewisville, TX 75029-2845

Business line: [972] 966-8135

Fax this form to: [972] 966-0219

Purchase online at our secure website: www.voicemailhell.com